DECIMALS: MULTIPLICATION AND DIVISION

Connie Eichhorn
Jean Rankin

Power Math Series

CAMBRIDGE Adult Education
A Division of Simon & Schuster
Upper Saddle River, New Jersey

Dr. Connie Eichhorn is the Supervisor of Transitional Programs for the Omaha Public Schools. She is the former president of the American Association of Adult and Continuing Education. Dr. Eichhorn is very active in adult education and has consulted in the development of a variety of adult education materials.

AUTHOR: Jean Rankin

EXECUTIVE EDITOR: Mark Moscowitz

EDITOR: Kirsten Richert

PRODUCTION DIRECTOR: Penny Gibson

PRODUCTION EDITOR: Linda Greenberg

PRINT BUYER: Patricia Alvarez

ART DIRECTOR: Marianne Frasco

BOOK DESIGN: Patrice Sheraton

ELECTRONIC PAGE PRODUCTION: Curriculum Concepts

COVER DESIGN: Amy Rosen

COVER PHOTO: © David Bishop/Phototake

Printed in the United States of America

1 2 3 4 5 6 7 8 9 10 99 98 97 96 95

ISBN 0-13-078866-X

CAMBRIDGE Adult Education
A Division of Simon & Schuster
Upper Saddle River, New Jersey

Contents

LESSON 1	Multiplying Whole Numbers	1
LESSON 2	Rounding to the Nearest Whole Number	3
LESSON 3	Rounding Off to Various Decimal Places	5
LESSON 4	Setting Up Multiplication Problems	7
LESSON 5	Multiplying a Whole Number by Tenths	9
LESSON 6	Multiplying a Whole Number by Hundredths	11
LESSON 7	Multiplying a Whole Number by Thousandths	13
LESSON 8	Multiplying a Whole Number by a Decimal	15
LESSON 9	Multiply a Decimal by a Decimal	17
LESSON 10	Zero as a Place Holder	19
LESSON 11	Multiplying Money	21
LESSON 12	Multiplying Small Decimal Numbers	23
LESSON 13	Multiplying a Decimal by 10, 100 and 1,000	25
LESSON 14	Rounding and Multiplying Decimals	27
LESSON 15	Rounding to the Leading Digit in Multiplication	29
LESSON 16	Review of Decimal Multiplication	31
LESSON 17	Review of Division of Whole Numbers	33
LESSON 18	Dividing a Decimal by a Whole Number	35
LESSON 19	Zero as a Place Holderin Division	38
LESSON 20	Dividing a Whole Number by a 10th, 100th, or 1,000th	41
LESSON 21	Dividing a Whole Number by a Decimal	43
LESSON 22	Dividing a Whole Number by a Larger Whole Number.	46
LESSON 23	Dividing a Decimal by a Decimal	48
LESSON 24	Dividing a Mixed Decimal by a Whole Number	50
LESSON 25	Dividing a Mixed Decimal by a Decimal	53
LESSON 26	Dividing a Mixed Decimal by a Mixed Decimal	55
LESSON 27	Dividing Money	57
LESSON 28	Dividing a Decimal by 10, 100 or 1,000	59
LESSON 29	Rounding to the Leading Digit in Division	61
LESSON 30	Decimal Division Review	63
LESSON 31	Review of Multiplication and Division of Decimals	65
APPENDIX:	Multiplication Facts	68
ANSWER KEY		69

To the Learner

The ten books in the Power Math series are designed to help you understand and practice arithmetic skills. Lessons are easy to use and the problems are designed to address every-day adult life.

Lessons have the following features:

- Every lesson begins with a sample problem from real-life experience. You are asked to use your knowledge of math to find a solution.
- The *Think* section takes you through the thought process you might use to organize the information in the problem and choose a problem-solving approach.
- The *Do* section shows you step-by-step how to solve the problem.
- In *Try These*, you will warm up by solving a few problems similar to the opening sample problem. Some steps are worked for you to get you off to a good start.
- The *Practice* section gives you ample opportunities to practice the skill presented in the lesson.
- The *Solving Problems* section applies the math skill in a practical application from your experience as a consumer and worker.

Within each book, review lessons give you opportunities to decide whether you have mastered the skills presented in the book. The *Answer Key* section at the end of the book has answers and worked-out solutions for the problems in the book. Use the answers to check your work. Use the worked-out solutions to make sure your approach to a problem was the correct one.

By working carefully through the exercises in this book, you will find increased confidence in your math skills. Good luck.

LESSON

1 Multiplying Whole Numbers

The dress shop that Ramiesha works for had a sale on women's coats.

Use the table below and find the total amount the store made from the sale of jackets.

Kind of Coat	Number Sold	Sale Price
Jackets	149	$ 36
Long Coats	115	$107
Leather Coat	70	$195

Think

To find the total amount, you need to multiply the number of jackets by the sale price of a jacket.

Do

Step 1. Set up the problem. Multiply by the digit in the ones place.

$$\begin{array}{r} 149 \\ \times\ 36 \\ \hline 894 \end{array}$$

$149 \times 6 = 894$

Step 2. Multiply by the 3 in tens place to find the next partial product.

$$\begin{array}{r} 149 \\ \times\ 36 \\ \hline 894 \\ 4470 \end{array}$$

When you multiply 149 by 3, you are really multiplying by 3 tens. Write a 0 in the ones place of the second product.

$139 \times 3 = 447$. Write the 7 in the tens place.

Step 3. Add the partial products and label your answer.

The amount of total sales from the jackets is $5,364.

$$\begin{array}{r} 149 \\ \times\ 36 \\ \hline 894 \\ 447 \\ \hline 5364 \end{array}$$

Try These

Find the product.

1. What was the total amount from the sale of long coats?

$$\begin{array}{r} 115 \\ \times\ 107 \\ \hline 805 \end{array} \qquad \begin{array}{r} 115 \\ \times\ 107 \\ \hline 805 \\ \_\ \_\ \_\ 0 \\ \hline \_\ \_\ \_\ \_\ \_\ \_ \end{array}$$

2. What was the total amount from the sale of leather coats? (**HINT:** Write the product of 195 × 7 to the left of the 0.)

$$\begin{array}{r} 195 \\ \times\ 70 \\ \hline 0 \end{array} \qquad \begin{array}{r} 195 \\ \times\ 70 \\ \hline \_\ \_\ \_\ \_\ 0 \end{array}$$

PRACTICE

Multiply.

3. 96 × 69 = _____
4. 341 × 204 = _____
5. 756 × 153 = _____
6. 164 × 628 = _____
7. 437 × 170 = _____
8. 1245 × 703 = _____
9. 562 × 812 = _____
10. 626 × 508 = _____
11. 3250 × 162 = _____
12. 836 × 480 = _____
13. 1893 × 414 = _____
14. 1071 × 376 = _____
15. 6410 × 209 = _____
16. 973 × 489 = _____
17. 23 × 4063 = _____

Solving Problems

Solve. A hailstorm hit Orem, Utah, and many roofs were damaged.

18. The Ace Roofing Company repaired 112 roofs, and it used an average of 37 squares for each. How many squares of shingles did Ace use?

____

19. Wong's Roofing repaired 130 roofs. It used an average of 206 squares of shingles per building. How many squares did Wong's use?

____

Check your answers on page 69.

LESSON

2 Rounding to the Nearest Whole Number

Amy needs to estimate the total weight of several packages. She rounds to the nearest whole number to make an estimate. Round the following weights to the nearest whole number.

Package 1	**2.41 lb.**
Package 2	**1.56 lb.**
Package 3	**3.27 lb.**

Think

Rounding off helps to estimate answers. The "nearest whole number" means the ones place.

Do

Step 1. Underline the number in the ones place.

Step 2. If the digit to its right is 5 or greater, add 1 to the underlined digit.

Step 3. If the digit to its right is less than 5, leave the underlined digit unchanged.

Package 1	2.41	4 is less than 5; 2.41 rounds to 2 pounds.
Package 2	1.56	5 is the number to the right of the underlined number. 1.56 rounds to 2 pounds.
Package 3	3.27	2 is less than 5. 3.27 rounds to 3 pounds.

Try These

Round to the nearest whole number.

1. 524.34   524.34   3 is less than 5   The rounded number is _____.

2. 370.76   370.76   7 is greater than 5   The rounded number is _____.

PRACTICE

Round to the nearest whole number.

3. 22.75 _____ **4.** 2.94 _____ **5.** 147.07 _____ **6.** 67.20 _____

7. 563.65 _____ **8.** 881.56 _____ **9.** 1342.26 _____ **10.** 369.4 _____

11. 7399.51 _____ **12.** 48.59 _____ **13.** 26.482 _____ **14.** 2487.593 _____

Solving Problems

Solve.

15. On a recent trip to visit Marina's family, the Bellas got 165.2 miles, 195.4 miles, and 183.7 miles on 3 tanks of gas. Round the amounts to the nearest whole number.

____

16. Write two numbers that round to 28 as the nearest whole number.

____

17. Write two numbers that round to 341 as the nearest whole number.

____

Check your answers on page 69.

LESSON

3 Rounding Off to Various Decimal Places

A mile is equal to 1609.344 meters. This number is rounded differently for various uses. Round this number to the nearest tenth, hundredth, and nearest whole number.

Where Used	Place to Round To	
Odometer in a car	Nearest tenth	1609.3̲44
Newspaper article	Nearest hundredth	1609.34̲4
Hikers Trail	Nearest whole number	1609̲.344

Think

To round a number, you must look at the digit to the right. If it is 5 or more, add 1 to the number. If it is 4 or less, leave the number. Then drop the remaining digits to the right.

Do

↓ 4 is less than 5

To the nearest tenth, 1609.3̲44 rounds to 1609.3.

↓ 4 is less than 5

To the nearest hundredth, 1609.34̲4 rounds to 1609.34.

↓ is less than 5

To the nearest whole number 1609̲.344 rounds to 1609.

Try These

Round as indicated.

1. Round 257.75 to the nearest tenth.

↓ the number to its right is 5.

257.7̲5

The number rounded to the nearest tenth is 257. _____.

2. Round 496.372 to the nearest hundredth.

↓ the number to its right is less than 5.

496.37̲2

The number rounded to the nearest hundredth is 496. _____.

PRACTICE

Round each of the following numbers to the underlined place value.

3. $473.2\underline{5}1$ _____ **4.** $56.\underline{3}85$ _____ **5.** $1235.\underline{0}99$ _____

6. $581.\underline{5}62$ _____ **7.** $96\underline{4}.53$ _____ **8.** $676.\underline{2}8$ _____

9. $47.56\underline{2}3$ _____ **10.** $29\underline{9}.53$ _____ **11.** $1.\underline{9}56$ _____

12. $.\underline{9}36$ _____

Solving Problems

Solve. The three NFL coaches with the highest career winning percentages are:

John Madden	**Raiders**	**.731**
Joe Gibbs	**Redskins**	**.683**
George Allen	**Rams, Redskins**	**.681**

13. Round these percentages to the nearest tenth. How does that affect the order of the list?

____

14. Round these percentages to the nearest hundredths. How does this change affect the order?

____

15. The 1991 team batting average for the Atlanta Braves was .258. The Los Angeles Dodgers average for the year was .253. Round these two averages to the nearest tenth.

____

Check your answers on page 69.

LESSON

4 Setting Up Multiplication Problems

Jade wants to multiply 51.37 by 4.1. How should she set up the problem. How many decimal points will there be in her answer?

Think

To multiply decimals, you need to write the numbers in a column. But you don't need to line up the decimal points. Instead, line up the last digits.

To figure out how many decimal places will be in the answer, count the number of decimal places in the problem.

Do

Step 1. Write the numbers in a column. Put the number with fewer digits on the bottom.

$$\begin{array}{r} 51.37 \\ \times \quad 4.1 \\ \hline \end{array}$$

Step 2. Count the decimal places in the problem. The answer will have 3 decimal places.

$$\begin{array}{r} 51.37 \\ \times \quad 4.1 \\ \hline \end{array} \qquad \begin{array}{r} \text{(2 places)} \\ + \text{ (1 place)} \\ \hline \text{(3 places)} \end{array}$$

Try These

Set up each multiplication problem. Determine the number of decimal places there will be in the answer.

1. 1392 × .871 =

$$\begin{array}{r} 1392 \\ \times \quad \underline{\qquad} \\ \hline \end{array}$$

There will be _____ decimal places in the answer.

2. 5.9 × .5831 =

$$\begin{array}{r} .5831 \\ \times \quad \underline{\qquad} \\ \hline \end{array}$$

There will be _____ decimal places in the answer.

PRACTICE

Rewrite each problem. Line up the last digits. How many decimal places will there be in each answer?

3. 1.392 × 1.51 =
Number of
decimal places _____

4. .52 × 1.79 =
Number of
decimal places _____

5. .427 × .81 =
Number of
decimal places _____

6. .385 × 2.9 =
Number of
decimal places _____

7. 11.074 × .8 =
Number of
decimal places _____

8. .917 × .18 =
Number of
decimal places _____

9. .97 × 1.01 =
Number of
decimal places _____

10. 6.84 × 3.98 =
Number of
decimal places _____

Solving Problems

Solve.

11. Which would be easier for you to multiply? Would the answers be the same?

A) 12407 × 356 or B) 356 × 12407

____

12. Susumu bought 1.57 lbs of fish for $1.69 a lb. How would you line up the numbers to multiply? How many decimal places would you expect to find in the answer?

____

Check your answers on page 69.

LESSON

5 Multiplying a Whole Number by Tenths

It rained .4 inches of rain on three consecutive days in Meridian. What was the total rainfall?

Think

You can multiply .4 by 3 or add .4 + .4 + .4.

Do

To multiply, line up the numbers, and multiply as though the numbers were whole numbers. Then place the decimal point.

$$\begin{array}{r} .4 \\ .4 \\ +\ .4 \\ \hline 1.2 \end{array} \qquad \begin{array}{r} .4 \\ \times\ 3 \\ \hline 1.2 \end{array}$$

The addition example shows how to place the decimal point in the answer. To place the decimal when multiplying, count the number of decimal places in the problem. There is one decimal place in the problem, so there is only one in the answer.

Try These

Multiply.

1. $\begin{array}{r} .7 \\ \times\ 4 \\ \hline \_\_.\_\_ \end{array}$ **2.** $\begin{array}{r} 9 \\ \times\ .3 \\ \hline \_\_.\_\_ \end{array}$ **3.** $\begin{array}{r} .7 \\ \times\ 1 \\ \hline \_\_.\_\_ \end{array}$ **4.** $\begin{array}{r} .2 \\ \times\ 0 \\ \hline \_\_.\_\_ \end{array}$

(HINT: You don't need to write the decimal point when zero is the answer.)

Multiply.

5. $\begin{array}{r} .6 \\ \times\ 3 \\ \hline \end{array}$

6. $\begin{array}{r} 4 \\ \times\ .8 \\ \hline \end{array}$

7. $\begin{array}{r} 9 \\ \times\ .6 \\ \hline \end{array}$

8. $\begin{array}{r} .6 \\ \times\ 1 \\ \hline \end{array}$

9. $\begin{array}{r} .7 \\ \times\ 9 \\ \hline \end{array}$

10. $\begin{array}{r} .3 \\ \times\ 5 \\ \hline \end{array}$

11. $\begin{array}{r} .9 \\ \times\ 0 \\ \hline \end{array}$

12. $\begin{array}{r} 5 \\ \times\ .5 \\ \hline \end{array}$

13. $\begin{array}{r} .6 \\ \times\ 7 \\ \hline \end{array}$

14. $\begin{array}{r} 8 \\ \times\ .5 \\ \hline \end{array}$

15. $\begin{array}{r} 9 \\ \times\ .4 \\ \hline \end{array}$

16. $\begin{array}{r} 3 \\ \times\ .7 \\ \hline \end{array}$

17. $\begin{array}{r} 6 \\ \times\ .9 \\ \hline \end{array}$

18. $\begin{array}{r} .9 \\ \times\ 9 \\ \hline \end{array}$

Solving Problems

Solve.

19. The 201 freeway has a callbox every .4 of a mile. Janice noticed she passes 6 callboxes on her way to work. How many miles does she drive on the 201 freeway?

____

20. At age 6, Becca has been growing an average of .3 centimeters every month for the past 4 months. How many centimeters has Becca grown in the past 4 months?

____

21. Bonita needs .8 gallons of gas to fill her small mower. If she used the mower 4 times in June, how much gas did she use?

____

22. It takes .3 pounds of hamburger to serve 1 person. Carlos wants to serve 9 people. How much hamburger will he need?

____

Check your answers on page 69.

LESSON

6 Multiplying a Whole Number by Hundreths

Brandon and his son ate three bags of peanuts at the baseball game. Each bag weighed .45 pounds. How many pounds of peanuts did they eat?

Think

Each bag weighed the same amount. You can multiply the weight of each bag by the number of bags.

Do

Step 1. Set up the problem. Since .45 contains more digits, put it on top.

$$\begin{array}{r} .45 \\ \times\quad 3 \\ \hline \end{array}$$

Step 2. Multiply as you would with whole numbers.

$$\begin{array}{r} .45 \\ \times\quad 3 \\ \hline 135 \end{array}$$

Step 3. Count the number of decimal places in the problem. There are two. Place the decimal point in the answer. Count two places from the right.

$$\begin{array}{r} .45 \\ \times\quad 3 \\ \hline 1.35 \end{array}$$

Brandon and his son ate 1.35 pounds of peanuts.

Try These

Multiply.

1. $\begin{array}{r} .81 \\ \times\quad 7 \\ \hline \end{array}$ **2.** $\begin{array}{r} .93 \\ \times\ 15 \\ \hline \end{array}$ **3.** $\begin{array}{r} .24 \\ \times\ 64 \\ \hline \end{array}$ **4.** $\begin{array}{r} .38 \\ \times\ 11 \\ \hline \end{array}$

Multiply.

5. $\begin{array}{r} 129 \\ \times\ .63 \\ \hline \end{array}$ **6.** $\begin{array}{r} 544 \\ \times\ .17 \\ \hline \end{array}$ **7.** $\begin{array}{r} 327 \\ \times\ .56 \\ \hline \end{array}$ **8.** $\begin{array}{r} 981 \\ \times\ .82 \\ \hline \end{array}$

9. $\begin{array}{r} .59 \\ \times\ 59 \\ \hline \end{array}$ **10.** $\begin{array}{r} 213 \\ \times\ .38 \\ \hline \end{array}$ **11.** $\begin{array}{r} 427 \\ \times\ .33 \\ \hline \end{array}$ **12.** $\begin{array}{r} 123 \\ \times\ .96 \\ \hline \end{array}$

13. 567 × .71 = _____ **14.** 951 × .59 = _____ **15.** 362 × .47 = _____

16. 424 × .31 = _____ **17.** 283 × .41 = _____ **18.** 730 × .93 = _____

19. .87 × 686 = _____ **20.** .32 × 341 = _____

Solving Problems

Solve.

21. The Rodriguez family is having a barbecue in the park on Labor Day. The Rodriguezes bought .75 pounds of chicken for each person. If there are going to be 12 people at the barbecue, how many pounds of chicken did the Rodriguezes buy?

_____

22. It is estimated that .62 parts of the population watch the news in the morning. Alliance has a population of 11,000. How many people in Alliance watch the news in the morning?

_____

23. Of the 950 people surveyed in Bentonville, .30 felt the local high school students should wear school uniforms. How many people wanted school uniforms?

_____

Check your answers on page 70.

LESSON

7 Multiplying a Whole Number by Thousandths

In a basketball tournament, a player averaged .375 points per minute played. If he played 32 minutes, how many points did he score?

Think

You need to multiply to find the number of points the player scored.

Do

Step 1. Set up the problem. Multiply as you would with whole numbers.

$$\begin{array}{r} .375 \\ \times \quad 32 \\ \hline 750 \\ 11250 \\ \hline 12000 \end{array}$$

Step 2. Count the number of decimal places in the problem. There are three. Count three places from the right to put the decimal point in the answer.

$$\begin{array}{r} .375 \\ \times \quad 32 \\ \hline 750 \\ 11\,250 \\ \hline 12.000 \end{array}$$

The player scored 12 points in 32 minutes.

Try These

Multiply.

1.
$$\begin{array}{r} .235 \\ \times \quad 21 \\ \hline 235 \\ 4\,700 \\ \hline 4.\_\_\_ \end{array}$$

2.
$$\begin{array}{r} .121 \\ \times \quad 48 \\ \hline \end{array}$$

3.
$$\begin{array}{r} .532 \\ \times \quad 13 \\ \hline \end{array}$$

Multiply.

4. $123 \times .451$

5. $634 \times .151$

6. $.732 \times 861$

7. $.983 \times 237$

8. $464 \times .366$

9. $.274 \times 37$

10. $.319 \times 72$

11. $.555 \times 18$

12. $.861 \times 84$

Solving Problems

Solve.

13. In 1993, Dale Jarrett won the Daytona 500 with an average speed of 154.972 miles per hour. How far had he driven at the end of three hours? How many miles did he have to drive to complete the race?

____

14. In 1906, Marriott set a one-mile speed record driving 127.659 miles per hour in a Stanley Steamer. How many miles could he have traveled in 3 hours if he had been able to maintain his record-setting speed?

____

Check your answers on page 70.

LESSON

8 Multiplying a Whole Number by a Decimal

After Mr. Ellis admitted that he rolled through a stop sign and hit Mrs. Lang's car, he was ordered to pay .8 of the $3,000 repair bill.

Think

Multiplying by a decimal is one way to find part of a number.

Do

Multiply: $3,000 × .8 =

Step 1. Set up the problem.

$$\begin{array}{r} 3000 \\ \times \quad .8 \\ \hline \end{array}$$

Step 2. Multiply as though the numbers were both whole numbers.

$$\begin{array}{r} 3000 \\ \times \quad .8 \\ \hline 24000 \end{array}$$

Step 3. Put the decimal point in the answer. Since there is one decimal point in the problem, count one decimal place from the right in the answer.

$$\begin{array}{r} 3000 \\ \times \quad .8 \\ \hline 2400.0 \end{array}$$

Mr. Ellis must pay $2,400.00 of the repair bill.

Try These

Multiply.

1. Find .01 of $250.

$$\begin{array}{r} 250 \\ \times \ .01 \\ \hline \end{array}$$

2. Multiply .051 by 6.

$$\begin{array}{r} .051 \\ \times \quad 6 \\ \hline \end{array}$$

3. Find .3 of 43.

$$\begin{array}{r} 43 \\ \times \ .3 \\ \hline \end{array}$$

Multiply.

4. 2.1×5

5. 3.9×2

6. $.04 \times 5$

7. $91 \times .2$

8. $11 \times .9$

9. $101 \times .05$

10. $82 \times .02$

11. $23 \times .03$

12. $.202 \times 6$

13. $.114 \times 2$

14. $.32 \times 3$

15. $92 \times .05$

Solving Problems

Solve.

16. Rocio walks 3.2 miles each day of the week. How many miles does she walk in 7 days?

____

17. Kim's Laundry uses 15 boxes of detergent each month. Each box weighs 9.64 kilograms. How much detergent is used in one month?

____

18. Harrison Waterworks hauls water to drilling rigs. One truck holds 3,200 gallons. One gallon of water weighs 8.36 pounds. What is the weight of the water?

____

19. The Chariot Inn received a shipment of 40 lamps. When the shipment was unpacked, .05 of the lamps were broken. How many lamps were broken?

____

Check your answers on page 70.

LESSON

9 Multiply a Decimal by a Decimal

Wanda's compact car gets 25.83 miles per gallon. How far can she drive on a tank of gas if her gas tank holds 14.6 gallons?

Think

You need to multiply the number of miles per gallon by the number of gallons in Wanda's gas tank.

Do

Step 1. Set up the problem. Multiply as you would with whole numbers.

```
    25.83
 ×   14.6
    15498
   10332
   2583
   377118
```

Step 2. Count the number of decimal places in the problem.

```
    25.83        2  decimal places
 ×   14.6      + 1  decimal place
   377118        3  decimal places
```

Step 3. Place the decimal point in the answer. Count three places from the right.

```
    25.83
 ×   14.6
  377.118
```

Wanda can drive 377.118 miles on a tank of gas.

Try These

Multiply.

1.
```
   1.45
 ×   .9
  _____
    .
```

2.
```
   34.2
 ×  .76
```

3.
```
   .812
 ×  5.4
```

4.
```
    .87
 ×  .31
```

PRACTICE

Place the decimal point correctly in the answer.

5. 4.26 × .51 = 21726

6. .781 × .3 = 2343

7. 25.14 × .76 = 191064

8. 123.7 × .14 = 17318

9. 6.2 × 3.7 = 2294

10. 1.7 × .8 = 136

11. .912 × .43 = 39216

12. 3.72 × 1.13 = 42036

Multiply.

13. 27.3 × .14

14. .832 × 7.5

15. 67.3 × 8.4

16. 7.9 × .61

17. 51.7 × .52 = _____

18. .465 × .67 = _____

19. .519 × .8 = _____

20. .39 × 7.16 = _____

Solving Problems

Solve.

21. A bottled drink contains .10 real fruit juice. One bottle of the drink contains .946 liters. How many liters of real juice are there in the bottle?

_____

22. Helen purchased a tablecloth that is .8 pure silk. How much silk is in the tablecloth if it weighs 11.2 ounces?

_____

Check your answers on page 70.

LESSON

10 Zero as a Place Holder

A machinest from Fitz Auto found a ball bearing in an assembly that was .005 centimeters too large. To bring it within tolerance, .2 of the .005 centimeters had to be removed. How much of the metal was removed?

Think

You need to multiply .005 by .2.

Do

Step 1. Set up the problem and multiply as you would with whole numbers.

```
  .005
×   .2
------
    10
```

Step 2. Count the decimal places in the problem. There must be four decimal places in your answer.

```
  .005        3  decimal places
×   .2      + 1  decimal place
------      -------------------
    10        4  decimal places
```

Step 3. The number of digits in your answer is less than four places. Begin at the right and count four places. Write zeros as place holders as needed.

```
  .005
×   .2
------
 .0010
```

You can drop the zero to the right of the 1 because it has no value.

The machinist had to remove .001 centimeter from the defective ball bearing.

Try These

Multiply.

1.
```
   .071   3  places
×   .35   2  places
-------
    355
   213
-------
  02485
```

2.
```
   .01
× .063
------
```

3.
```
  0.546
×   .82
-------
```

Multiply.

4. $\begin{array}{r} .093 \\ \times\ .65 \\ \hline \end{array}$

5. $\begin{array}{r} .58 \\ \times\ .078 \\ \hline \end{array}$

6. $\begin{array}{r} 3.412 \\ \times\ .008 \\ \hline \end{array}$

7. $\begin{array}{r} .003 \\ \times\ .42 \\ \hline \end{array}$

8. $\begin{array}{r} .093 \\ \times\ .1 \\ \hline \end{array}$

9. $\begin{array}{r} 1.342 \\ \times\ .063 \\ \hline \end{array}$

10. $\begin{array}{r} .924 \\ \times\ .038 \\ \hline \end{array}$

11. $\begin{array}{r} .26 \\ \times\ .04 \\ \hline \end{array}$

12. $\begin{array}{r} 1.56 \\ \times\ .03 \\ \hline \end{array}$

13. $\begin{array}{r} .741 \\ \times\ .053 \\ \hline \end{array}$

14. $\begin{array}{r} 9.3 \\ \times\ .009 \\ \hline \end{array}$

15. $\begin{array}{r} .12 \\ \times\ .12 \\ \hline \end{array}$

Solving Problems

Solve.

16. One-tenth of the people involved in a health study were found to have a weight problem. Of this group, .35 had a heart problem. What part of the people in the study had a weight and heart problem? (**HINT:** You need to multiply .1 by .35 to find the answer.)

____

17. By May of this year, only .4 of the members of a union had paid their annual dues. About .2 of that group had attended meetings. What part of the total membership had not paid dues but had attended meetings?

____

Check your answers on page 70.

LESSON

11 Multiplying Money

Mrs. Chim bought a 12-pound turkey for Thanksgiving dinner. Turkey costs 89 cents per pound. How much did she have to pay?

Think

You need to multiply 89 cents by 12. Eighty-nine cents is 89 hundredths of a dollar or $.89.

Do

Step 1. Set up the problem. Write the cents as a decimal. Then multiply as you would with whole numbers.

```
   .89
×   12
   178
   89
  1068
```

Step 2. There are two decimal places in the problem. Count two digits from the right in the answer and put in the decimal point.

```
   .89
×   12
 10.68
```

Step 3. Write the answer in dollars and cents.

10.68 = $10.68

Mrs. Chim paid $10.68 for the turkey.

Try These

Multiply.

1.
```
  $7.18
×    34
   2872
  2154
```

2.
```
  $6.11
×    16
```

3.
```
     591
× $  .42
```

Multiply.

4. 531 × $ .26

5. 283 × $ .44

6. $1.95 × 68

7. 733 × $1.17

8. $6.29 × 18

9. $1.96 × 83

10. $4.76 × 91

11. $8.42 × 18

Solving Problems

Solve. Derek serviced a car for a family going on a vacation. He bought the following products from Auto Serv so that he could do the work.

AUTO SERV Order

Quantity	Description	Unit Cost	Total
5	Ozark Motor Oil	1.15	
1	Ozark Oil Filter	3.78	
6	Better Spark Plugs	2.73	
2	Barr's Lube (Tube)	1.98	
2	Ozark Coolant	6.94	
		Total	

12. Complete the table. Multiply the quantity of each item by the unit price to find the cost of that item. Then add the amounts for the items to find the total.

13. If Auto-Serv charged .1 of the total cost to deliver his order, what was the cost of delivery?

Check your answers on pages 70–71.

LESSON

12 Multiplying Small Decimal Numbers

About .01 of all individuals who pay federal income tax are audited. Of this group, .1 will have money returned to them. What part of those who file will receive a refund?

Think

To find the part that Internal Revenue owes money to, you can multiply .01 by .1.

Do

Step 1. Set up the problem. Multiply as you would whole numbers.

$$\begin{array}{r} .01 \\ \times \quad .1 \\ \hline 1 \end{array}$$

Step 2. Count the decimal places in the problem. Then put the decimal point in the answer.

$$\begin{array}{rl} .01 & \quad 2 \text{ decimal places} \\ \times \quad .1 & \quad +\ 1 \text{ decimal place} \\ \hline .001 & \quad 3 \text{ decimal places} \end{array}$$

Of those who file an income tax report and are audited, .001 will receive a refund. In other words, the government owes a refund to one person out of each thousand who are audited.

Try These

Multiply.

1. $\begin{array}{r} .15 \\ \times\ .05 \\ \hline .\_\_\_\_ \end{array}$ **2.** $\begin{array}{r} .26 \\ \times\ \ .1 \\ \hline .\_\_\_ \end{array}$ **3.** $\begin{array}{r} .031 \\ \times\ \ .87 \\ \hline \end{array}$ **4.** $\begin{array}{r} .6 \\ \times\ .04 \\ \hline \end{array}$

Multiply.

5. $\begin{array}{r} .42 \\ \times\ .01 \\ \hline \end{array}$	**6.** $\begin{array}{r} 5.7 \\ \times\ .13 \\ \hline \end{array}$	**7.** $\begin{array}{r} .2 \\ \times\ .7 \\ \hline \end{array}$	**8.** $\begin{array}{r} .15 \\ \times\ .33 \\ \hline \end{array}$
9. $\begin{array}{r} .81 \\ \times\ .7 \\ \hline \end{array}$	**10.** $\begin{array}{r} .63 \\ \times\ 10 \\ \hline \end{array}$	**11.** $\begin{array}{r} .05 \\ \times\ .1 \\ \hline \end{array}$	**12.** $\begin{array}{r} .84 \\ \times\ .42 \\ \hline \end{array}$
13. $\begin{array}{r} .29 \\ \times\ 1.7 \\ \hline \end{array}$	**14.** $\begin{array}{r} .071 \\ \times\ .13 \\ \hline \end{array}$	**15.** $\begin{array}{r} .38 \\ \times\ .22 \\ \hline \end{array}$	**16.** $\begin{array}{r} 3.4 \\ \times\ .27 \\ \hline \end{array}$

Solving Problems

Solve.

17. A truck has a load limit of .75 tons. If the truck is .5 full, how many pounds is the truck carrying. (**HINT:** 1 ton = 2,000 lbs)

____

18. One centimeter is about .39 of an inch. Five-tenths of a centimeter is what part of an inch?

____

19. In a driver's education class, .65 of the students completed the course successfully. Of those, .4 passed the test at the DMV. What part of those who took the course went on to obtain a driver's license?

____

20. An eight ounce serving of orange juice will supply .08 of the recommended daily allowance (RDA) of Vitamin A. Elaine drank .5 of a serving of orange juice. What part of the RDA for Vitamin A did she get from the juice?

____

Check your answers on page 71.

LESSON

13 Multiplying a Decimal by 10, 100 and 1,000

Nicholas lives 3.8 miles from the bank where he works. He has decided to ride his bike to and from work every day to save money on bus fares. How many miles will he ride in a five-day week?

Think

Monday through Friday is five days, and two trips a day is ten trips. You need to multiply 3.8 by 10.

Do

Set up the problem. Then multiply as you would whole numbers. Put the decimal point in the answer.

$$\begin{array}{r l} 3.8 & \text{1 decimal place} \\ \times\ \ 10 & \text{0 decimal places} \\ \hline 38.0 & \end{array}$$

Nicholas will ride his bike 38 miles in a week.

Now compare the number you started with and the answer. What do you notice? The decimal point in 3.8 moved one place to the right.

You can use a shortcut whenever you multiply by 10, 100, or 1,000. Move the decimal point to the right one decimal place for each zero in the number you are multiplying by.

To multiply by 10, move the decimal point 1 place to the right.

To multiply by 100, move the decimal point 2 places to the right.

To multiply by 1,000, move the decimal point 3 places to the right.

Examples:

$$\begin{array}{r} 1.98 \\ \times\ \ 10 \\ \hline 19.\underline{8} \end{array} \qquad \begin{array}{r} 1.98 \\ \times\ 100 \\ \hline 1\underline{98}. \end{array} \qquad \begin{array}{r} 1.98 \\ \times\ 1000 \\ \hline 1\underline{980}. \end{array} \qquad \begin{array}{r} .037 \\ \times\ 100 \\ \hline 3.7 \end{array}$$

Try These

Multiply.

1. $\begin{array}{r} .045 \\ \times \quad 10 \\ \hline \end{array}$

2. $\begin{array}{r} 1.78 \\ \times \quad 100 \\ \hline \end{array}$

3. $\begin{array}{r} 2.39 \\ \times \quad 100 \\ \hline \end{array}$

4. $\begin{array}{r} .517 \\ \times \quad 1000 \\ \hline \end{array}$

PRACTICE

Multiply.

5. 5.3 × 10 = _____

6. 178.9 × 100 = _____

7. 6.12 × 100 = _____

8. .34 × 1000 = _____

9. 24.71 × 1000 = _____

10. 1.6 × 100 = _____

11. .067 × 10 = _____

12. 1.517 × 100 = _____

13. 8.03 × 10 = _____

14. 2.357 × 1000 = _____

Solving Problems

Solve.

15. Last week Ramón worked 10 hours for Highland Cleans at $10 per hour. He also worked 10 hours for Plummer's Furniture at $7.50 per hour. How much did he earn from each part-time job last week?

_____

16. Studies show that the average person is responsible for about 4.38 pounds of trash per day. How many pounds per day of trash would a community of 1,000 people throw away?

_____

Check your answers on page 71.

LESSON

14 Rounding and Multiplying Decimals

The Domanski family bought 26.5 gallons of gas on their trip to visit their grandmother. The average price per gallon was $1.29. To the nearest dollar, about how much did they spend on gas?

Think

To solve the problem, you need to multiply the number of gallons by the cost per gallon. Since you only need an estimate, round the numbers before multiplying.

Do

Step 1. Round the numbers as shown.

26.5 rounds to 27 gallons

$1.29 rounds to $1.30

Step 2. Multiply.

$$\begin{array}{r} \$1.30 \\ \times \quad 27 \\ \hline 35.10 \end{array}$$

Step 3. Round to the nearest dollar.

$35.10 rounds to $35.

The family spent about $35 on gas for their trip.

Try These

Estimate an answer. Round and multiply.

1. $\begin{array}{r} 5.8 \\ \times \ .17 \\ \hline \end{array}$ 5.8 is closer to 6 than 5; .17 is closer to .2 than .1 $\begin{array}{r} 6 \\ \times \ .2 \\ \hline 1.2 \end{array}$

2. $\begin{array}{r} \$6.98 \\ \times \quad .17 \\ \hline \end{array}$ $6.98 is close to $7; .17 is closer to _____

3. $\begin{array}{r} 4.67 \\ \times \ .95 \\ \hline \end{array}$

PRACTICE

Choose the best method to estimate an answer.

4. $.15 \times .9 =$	**a.** $.2 \times 1$	**b.** 0×1	**c.** $.2 \times .9$
5. $7.85 \times .32 =$	**a.** $8 \times .3$	**b.** 8×1	**c.** $7.9 \times .4$
6. $11.36 \times .95 =$	**a.** $12 \times 1 =$	**b.** 11×1	**c.** $12 \times .99$
7. $178.5 \times .52 =$	**a.** $170 \times .5$	**b.** $179 \times .6$	**c.** $179 \times .5$
8. $96.4 \times 1.67 =$	**a.** 96×1.7	**b.** 96.5×1.7	**c.** 97×2
9. $9{,}367.5 \times 20.9 =$	**a.** $10{,}000 \times 20$	**b.** $9{,}400 \times 21$	**c.** $9{,}300 \times 20$

Round these numbers to the nearest whole number and multiply.

10. $\begin{array}{r} 58.6 \\ \times \quad 4.9 \\ \hline \end{array}$

11. $\begin{array}{r} 610.7 \\ \times \quad 19.6 \\ \hline \end{array}$

12. $\begin{array}{r} 1.29 \\ \times \quad 1.6 \\ \hline \end{array}$

13. $\begin{array}{r} 13.40 \\ \times \quad 76.9 \\ \hline \end{array}$

14. $\begin{array}{r} 47.1 \\ \times \quad 9.8 \\ \hline \end{array}$

15. $\begin{array}{r} 3.9 \\ \times \quad 2.09 \\ \hline \end{array}$

Solving Problems

Solve.

16. Use the formula: The weight of an unrestrained person × speed of car = the force of impact in pounds.

Estimate the impact force of an unrestrained person who weighs 146.5 pounds and is thrown from a car traveling 53.5 miles an hour?

____

17. Cameron, who works as a tour guide in Wyoming, has a summer salary of $7,325.50. He estimates that .27 of his salary is spent on housing. Estimate the cost of his housing for the summer.

____

Check your answers on page 71.

LESSON

15 Rounding to the Leading Digit in Multiplication

Ron Hunter drove .746 hours at a rate of 251.6 mi/hour. About how many miles did he travel?

Think

The leading digit of any number is the first digit on the left.

One quick way to estimate an answer is to round the numbers in the problem to their leading digits and multiply. To solve this problem, you need to round .746 and 251.6 to their leading digits.

Do

Step 1. Round the numbers in the problem.

251.6	2 is the leading digit; 5 is the number to its right. 251.6 rounds to 300.
.746	7 is the leading digit; 4 is the number to its right. 4 is less than 5 so .746 rounds to .7.

Step 2. Multiply using the rounding numbers.

$$\begin{array}{r} 300 \\ \times\ \ .7 \\ \hline 210.0 \end{array}$$

Ron Hunter drove about 210 miles.

Try These

Round to the leading digits and multiply.

1. $\begin{array}{r} .356 \\ \times\ .23 \\ \hline \end{array}$ ⟶ $\begin{array}{r} .4 \\ \times\ .2 \\ \hline \end{array}$._ _

2. $\begin{array}{r} 48.9 \\ \times\ \ .92 \\ \hline \end{array}$

3. $\begin{array}{r} 9.56 \\ \times\ \ 5.6 \\ \hline \end{array}$

PRACTICE

Round to the leading digits and multiply.

4. .751 × .69 = _____

5. 1.75 × .35 = _____

6. 304.7 × .08 = _____

7. 93.4 × 5.6 = _____

8. 617.5 × 3.46 = _____

9. 593.1 × .046 = _____

10. 44.31 × .119 = _____

11. .273 × .56 = _____

12. 1,392.1 × 1.631 = _____

13. 5,280 × .56 = _____

14. 889.1 × 9.9 = _____

15. 77.8 × .67 = _____

Solving Problem

Solve. First, estimate answers by rounding to the leading digit. Then solve the problems to find the exact answers.

16. Liane bought 7.85 pounds of peaches. Each pound cost 59.5 cents. Find the cost of the peaches.

____

17. Ridgecrest School charges $30.25 per credit hour for evening classes. Sergio has registered for a 2.5-hour class to prepare for the GED test. What will he be charged for the course?

____

18. Of the 115.3 million taxpayers that file federal tax returns each year, about .05 ask for an extension of time to file a return. How many taxpayers ask for an extension?

____

Check your answers on page 72.

LESSON

16

Review of Decimal Multiplication

Multiply.

1.	$25.50 × 100	**2.**	.16 × .5	**3.**	.006 × 12	**4.**	500 × .25
5.	112.8 × 4.9	**6.**	.009 × .1	**7.**	$10.75 × 1000	**8.**	4.5 × .6
9.	3.708 × .41	**10.**	.006 × .394	**11.**	5.37 × 19	**12.**	.5768 × 1000
13.	7.18 × 2.26	**14.**	13.76 × 520	**15.**	93.6 × 42.7	**16.**	605.1 × 6.18
17.	.35 × .25	**18.**	73.61 × 1.01	**19.**	8.28 × .17	**20.**	65.24 × .01

Solving Problems

Solve.

21. Twenty-six boys went on an overnight camping trip with their church group. The church paid for the transportation, meals, and camping fees. These totaled $30.45 per person. What was the total amount the church paid for the boys to go on the trip?

____

22. Javier and Stephanie bought out-of-state fishing licenses. Each license was $20 and the two trout stamps were $4.50 each. What was the total cost of their licenses and stamps?

____

23. Li-Young bought five autographed baseball cards. Three of the cards sold for $18.95 each. The other two cards cost $29.56. Estimate the total cost of the cards.

____

24. For one city the 1992 census reported a population of 180,560 with an expected growth rate of .542 over the next 10 years. How many additional people should the city plan to serve 10 years from now? Round to the nearest whole number.

____

Check your answers on page 72.

LESSON

17 Review of Division of Whole Numbers

A shipment of fishing lures arrived at Odion Sports Supply packaged 5,000 lures to a box. Etty's job is to repackage them in plastic bags of 12 each. How many packages can she make from each box of 5,000. Were there any left over?

Think

You need to divide 5,000 by 12. If there is a remainder, those are the ones that are left over.

Do

Step 1. Set up the problem.

Step 2. Estimate how many times 12 will go into 50. It will go 4 times.

Step 3. Subtract: 50 – 48 = 2. Bring down the 0.

$$\begin{array}{r} 4 \\ 12\overline{)5000} \\ \underline{48} \\ 20 \end{array}$$

Step 4. Continue dividing. When you have brought down all of the digits you are finished. You have a remainder of 8.

$$\begin{array}{r} 416 \\ 12\overline{)5000} \\ \underline{48} \\ 20 \\ \underline{12} \\ 80 \\ \underline{72} \\ 8 \end{array}$$

Etty can make 416 packages from a box of 5,000 lures. There will be eight left over.

Try These

Divide.

1. $\begin{array}{r} 2 \\ 3\overline{)726} \\ \underline{6} \\ 12 \end{array}$

2. $4\overline{)6480}$

3. $5\overline{)245}$

Divide. Show any remainder.

4. $8\overline{)4808}$ **5.** $15\overline{)625}$ **6.** $73\overline{)5321}$

7. $57\overline{)1029}$ **8.** $142\overline{)2976}$ **9.** $9\overline{)1358}$

10. $510\overline{)37230}$ **11.** $36\overline{)1260}$ **12.** $342\overline{)43{,}776}$

Solving Problems

Solve.

13. Phillip ordered 102 rose bushes. He selected an equal number of six different varieties. How many of each variety did he order?

____

14. On her 2,450-mile trip to visit her sister, Inez intends to drive the same number of miles each day. She plans to finish the trip in seven days. How many miles should she plan on driving each day?

____

Check your answers on page 72.

LESSON

18 Dividing a Decimal by a Whole Number

La Vonne bought .75 of a pound of caramel candy at Miller's Candy Store. She divided it evenly among her 3 children. How much did each child receive?

Think

You need to divide the amount of candy by the number of children. Since .75 pound is less than 1 pound, you know each child will receive less than a pound. Your answer will be a decimal number.

Do

Step 1. Set up the problem. $3\overline{).75}$

↑ divisor

Step 2. Since the divisor is a whole number, place the decimal point directly above the decimal point in the number to be divided. $3\overline{).75}$ (decimal point placed above)

Step 3. Divide as you would with whole numbers.

```
   .25
3).75
   6
   15
   15
```

Each of La Vonne's children received .25 lb of candy.

Try These

Divide.

1. $42\overline{)109.2}$ **2.** $16\overline{)59.2}$ **3.** $8\overline{)9.98}$

Divide.

4. $47\overline{)4.794}$ **5.** $35\overline{)12.25}$ **6.** $6\overline{)106.2}$

7. $18\overline{)8.28}$ **8.** $29\overline{)12.76}$ **9.** $51\overline{)183.6}$

10. $76\overline{)4.180}$ **11.** $9\overline{)10.80}$ **12.** $84\overline{)5.040}$

13. $42\overline{)466.2}$ **14.** $19\overline{)1.368}$ **15.** $59\overline{)35.99}$

Solving Problems

Solve.

16. The Adams and Son Trucking Company will move the Cortez Drilling Company rig from Ogden, Utah, to Sterling, Colorado. The total weight is 260.4 tons. Each truck has the capacity to haul 22 tons. How many trucks will they need to do the job?

____

17. Will the trucking company need a truck to carry part of a load?

____

18. Carmen spent $529.44 for electricity in 1993. What was the average cost per month for the year?

____

19. Maya drove her car 330.6 miles in 6 hours. How many miles did she travel in one hour?

____

20. Mr. De Vera sent all of his five grandchildren a year's subscription to a children's magazine for their birthdays. The total cost was $99.75. What was the cost of each subscription?

____

Check your answers on page 72.

LESSON

19 Zero as a Place Holder in Division

Diamond Land Jewelers has an order to make five identical platinum rings. The jeweler has .315 ounces of platinum to use for the rings. How much can he use for each ring?

Think

Divide the amount of platinum by the number of rings to find the amount in one ring.

Do

Step 1. Place the decimal point above the decimal point in the number you are dividing.

$$5\overline{).315}$$ (with the decimal point placed above)

Step 2. You cannot divide 5 into 3, so you need to place a zero over the 3 to hold the decimal point in the correct position.

$$\begin{array}{r} .0 \\ 5\overline{).315} \end{array}$$

Step 3. Continue your division as you did with whole numbers.

$$\begin{array}{r} .063 \\ 5\overline{).315} \\ \underline{30} \\ 15 \\ \underline{15} \end{array}$$

Each ring will contain .063 ounces of platinum.

Try These

Divide.

1. $\begin{array}{r} .0 \\ 6\overline{).546} \end{array}$ **2.** $45\overline{)3.240}$ **3.** $87\overline{).7917}$

PRACTICE

Divide.

4. $3\overline{)\,.12}$

5. $14\overline{)\,.70}$

6. $25\overline{)\,.775}$

7. $14\overline{)\,.7014}$

8. $132\overline{)\,11.88}$

9. $256\overline{)\,18.688}$

10. $33\overline{)\,.627}$

11. $63\overline{)\,10.71}$

12. $56\overline{)\,6.048}$

13. $94\overline{)\,1.786}$

14. $476\overline{)\,3.8556}$

15. $281\overline{)\,1.6768}$

Solving Problems

Divide.

16. Jonas bought two identical scarves for $6.10. What was the cost of each?

____

17. Marisol and Kim rode a shuttle bus to the aiport. The total cost for both riders was $22.12. What was each rider's share?

____

18. Erin cut a board 16.08 feet long into four pieces to put shelves in his bedroom closet. Disregarding the waste caused by the saw, what was the length of each piece?

____

Check your answers on page 72.

LESSON

20 Dividing a Whole Number by a 10th, 100th, or 1,000th

A millimeter is .1 of a centimeter. How many millimeters are there in 4 centimeters?

Think

Divide 4 centimeters by the number of millimeters in 1 centimeter.

You will be dividing a whole number by a decimal.

Do

Step 1. Set up the problem. Write a decimal point after the whole number. $.1\overline{)4.}$

Step 2. Move the decimal point in the divisor one place to the right to make the divisor a whole number $1.\overline{)4.}$

Step 3. Move the decimal point in the dividend one place to the right. Use 0 as a place holder. $1.\overline{)40.}$

Step 4. Divide as you would with whole numbers. $\begin{array}{r} 40 \\ 1\overline{)40} \end{array}$

There are 40 millimeters in 4 centimeters.

Look at the number fact: 4 ÷ .1 = 40. Dividing by .1 was the same as multiplying the dividend by 10.

To divide by .1, multiply the dividend by 10.

To divide by .01, multiply the dividend by 100.

To divide by .001, multiply the dividend by 1,000.

Try These

Divide.

1. $.01\overline{)7}$ $\quad 01.\overline{)7.00.}$ **2.** $.001.\overline{)42.\_\,\_\,\_.}$

Each time you moved the decimal point the number of places that were needed to make the divisor a whole number.

PRACTICE

Divide.

3. $.1\overline{)320}$ **4.** $.1\overline{)9}$ **5.** $.01\overline{)9}$ **6.** $.001\overline{)74}$

7. $.01\overline{)54}$ **8.** $.001\overline{)11}$ **9.** $.1\overline{)65}$ **10.** $.01\overline{)121}$

11. $.001\overline{)26}$ **12.** $.1\overline{)1}$ **13.** $.001\overline{)93}$ **14.** $.1\overline{)411}$

15. $.01\overline{)364}$ **16.** $.001\overline{)701}$ **17.** $.01\overline{)10}$ **18.** $.1\overline{)24}$

Solving Problems

Solve.

19. Only six students failed a math test. If this represents .1 of the total class, how many were in the class?

____

20. One business found that .01 of their employees traveled over 50 miles to work. If 29 workers traveled over 50 miles to work, how many total employees work for the company?

____

Check your answers on page 72.

LESSON

21 Dividing a Whole Number by a Decimal

Meaghan built birdhouses for Christmas presents. She bought 9 feet of dowel rod for the perches and divided it into sections .5 feet long. How many perches did she have?

Think

Divide the total length by the length of one perch to find the number of perches.

Do

Step 1. Set up the problem. You need to divide a whole number by a decimal. $.5\overline{)9}$

Step 2. Place the decimal after the whole number. $.5\overline{)9.}$

Step 3. Move the decimal point in the divisor one place to the right. You need to make the divisor a whole number. $5.\overline{)9.}$

Step 4. Move the decimal point in the dividend one place to the right. You will need to add a zero to use as a place holder. $.5\overline{)9.0}$

Step 5. Divide as you would with whole numbers. You do not need to place a decimal after 18, because 18 is a whole number.

$$\begin{array}{r} 18. \\ 5.\overline{)90.} \\ \underline{5} \\ 40 \\ \underline{40} \end{array}$$

Meaghan cut 18 perches from the dowel rod.

Try These

Divide.

1. $.6\overline{)30}$ **2.** $.45\overline{)45}$ **3.** $.125\overline{)75}$

Divide.

4. $.31\overline{)124}$ **5.** $.22\overline{)11}$ **6.** $.134\overline{)804}$ **7.** $.7\overline{)119}$

8. $.64\overline{)1088}$ **9.** $.09\overline{)1917}$ **10.** $.43\overline{)2881}$ **11.** $.85\overline{)7225}$

12. $.5\overline{)515}$ **13.** $.073\overline{)6278}$ **14.** $.91\overline{)5005}$ **15.** $.28\overline{)112}$

Solving Problems

Solve.

16. The Huang family is planting a border for a flower bed that is 30 feet long. The directions suggest spacing the daffodils .75 feet apart. How many bulbs will they need to plant the 30 feet of border?

____

17. A relay team ran 8 kilometers on a .4 kilometer track. How many laps did the team run?

____

18. The Goldman Bakery mixed 9 pounds of fruitcake batter. It was divided into .9 pound amounts and baked. How many individual cakes did they bake?

____

19. A low-fat granola bar weighs 2.75 oz. and contains 132 calories. Find the number of calories per ounce.

____

Check your answers on page 73.

LESSON

22 Dividing a Whole Number by a Larger Whole Number

Angela has been hired to cook for a dinner party. She bought 4 pounds of sirloin steak at the meat market. She plans on serving 10 people at the dinner. What will be the size of each serving?

Think

Divide the amount of food by the number of servings. Since there are more servings than there are pounds of sirloin, you know that each serving will be less than a pound.

Do

Step 1. Set up the problem. Place a decimal point after the 4. $10\overline{)4.}$

Step 2. Place a decimal point above the decimal point in the dividend. $10\overline{)4.}$ (with . above)

Step 3. Since you cannot divide 4 by 10, add a 0 after the decimal point and divide as you would with whole numbers. $\begin{array}{r} .4 \\ 10\overline{)4.0} \end{array}$

Each person will receive a .4 pound serving of meat.

Try These

Divide.

1. $140\overline{)105.}$

2. $152\overline{)133.000}$ (with . above)

3. $\begin{array}{r} .666 \\ 618\overline{)412.000} \\ \underline{370\ 8} \\ 41\ 20 \\ \underline{37\ 08} \\ 4\ 120 \end{array}$

In problem 3, the answer is a repeating decimal. The 6 will keep repeating. Once a decimal starts to repeat, you can show a remainder or round the decimal to an appropriate place value.

$$\begin{array}{r} .666 \\ 618\overline{)412.000} \end{array}$$

Round the answer to the nearest hundredth.

.666

6 is in the hundredths place, the 6 to its right is more than 5.

The rounded number is _____.

PRACTICE

Divide. Carry division out to the thousandths place and then round to the hundredths.

4. $252\overline{)104.000}$

5. $92\overline{)56}$

6. $105\overline{)5}$

7. $462\overline{)291}$

8. $9\overline{)2}$

9. $624\overline{)84}$

Divide and round the quotient to the nearest thousandth.

10. $321\overline{)126}$

11. $84\overline{)4}$

12. $78\overline{)63}$

Solving Problems

Solve.

13. The Rockies' first baseman had 65 hits out of 195 times at bat. What is his average? (HINT: Divide the number of hits by the number of times at bat.) Express your answer to the nearest thousandth.

____

14. The Rockies won 39 games out of 81 games played at home. Divide the games won by the total games played at home to find their winning percentage at home. Express your answer to the nearest thousandth.

____

Check your answers on page 73.

LESSON

23 Dividing a Decimal by a Decimal

Rakim bought .9 tons of fertilizer for the nursery. He plans to use .15 tons for each bed. How many beds can be fertilized?

Think

Divide the amount of fertilizer by the amount for each bed.

Do

Step 1. Set up the problem. $.15\overline{).9}$

Step 2. Move the decimal point two places to right to make the divisor a whole number. $15.\overline{).9}$

Step 3. Move the decimal point the same number of places in the dividend. You will have to add a zero. $15.\overline{)90.}$

Step 4. Place the decimal point in the quotient directly above the one in the dividend. $15.\overline{)90.}$

Step 5. Divide as you would with whole numbers.

$$\begin{array}{r} 6 \\ 15\overline{)90.} \\ \underline{90} \end{array}$$

Rakim can fertilize 6 beds at the nursery.

Try These

Divide.

1. $.17\overline{).68}$

2. $.125\overline{).1}$

3. $.471\overline{).10833}$

PRACTICE

Divide.

4. $.22\overline{).2156}$

5. $.81\overline{).6156}$

6. $.13\overline{).169}$

7. $.05\overline{).55}$

8. $.416\overline{).3744}$

9. $.26\overline{).0273}$

10. $.19\overline{).1577}$

11. $.265\overline{).2915}$

12. $.07\overline{).049}$

13. $.75\overline{).5625}$

14. $.004\overline{).4144}$

15. $.819\overline{).37674}$

Solving Problems

Solve.

16. Miriam has $.93 in her pocket. She wants to buy stickers to use as rewards for the preschool children she teaches. If the stickers are $.04 each, how many can she buy? Round answer down to the nearest whole number.

____

17. Pardee mixed .976 cubic yards of concrete to make stepping stones for a path. Each stone uses .0122 cubic yards. How many stones could he make? Will he have any concrete left over?

____

Check your answers on page 73.

LESSON

24 Dividing a Mixed Decimal by a Whole Number

During a scout hike, Tyrell walked 18.3 miles in 6 hours. What was his rate per hour?

Think

You need to divide the distance by the time.

Do

Step 1. Set up the problem. $6\overline{)18.3}$

Step 2. Place the decimal point above the decimal point in the dividend. $\overset{.}{6\overline{)18.3}}$

Step 3. Divide as you would with whole numbers. Add zeros to the right of the last decimal digit to continue the division.

$$\begin{array}{r} 3.15 \\ 6\overline{)18.30} \\ \underline{18} \\ 30 \\ \underline{30} \end{array}$$

Tyrell walked 3.15 miles per hour.

Suppose you need to find this answer to the nearest tenth. You would still need to carry out the division to the hundredths place. Always continue dividing until you have one place more than the place you need to round to.

For example, to round to the hundredths place, carry the division out to the thousandths place.

Try These

Divide.

1. $75\overline{).225}$ **2.** $412\overline{)49.44}$ **3.** $92\overline{)27.027}$

PRACTICE

Divide. Round each answer to the nearest tenth.

4. $5\overline{)4.05}$

5. $12\overline{)2.58}$

6. $7\overline{)1.414}$

7. $4\overline{)10.24}$

8. $9\overline{)7.14}$

9. $10\overline{)5.06}$

Divide. Round these answers to the nearest hundredth.

10. $42\overline{)6.930}$

11. $67\overline{)7.437}$

12. $51\overline{)15.606}$

13. $25\overline{)1.85}$

14. $9\overline{)1.5}$

15. $6\overline{)1.452}$

Divide. Round these answers to the nearest thousandth.

16. $16\overline{)10.2}$

17. $7\overline{)3.02}$

Solving Problems

Solve.

18. Eleda mailed six identical packages to her grandchildren in Houston. The total weight was 25.2 pounds. What was the weight of each package?

____

19. Leonisa bought a remnant of drapery material at Yardage Shack. The remnant is 40.2 yards long. She wants to drape six windows. How many yards will she have for each window?

____

Check your answers on page 73.

LESSON

25 Dividing a Mixed Decimal by a Decimal

Radhika is cooking a ham. Her cookbook states that a ham should bake 20 minutes per pound. Radhika doesn't know what the ham weighs, but she remembers that it was on sale for \$.95 a pound and the total cost was \$15.35. What did the ham weigh to the nearest tenth of a pound?

Think

Divide the total cost of the ham by the cost per pound. To find the weight to the nearest tenth of a pound, you will need to carry out the division to the hundredths place.

Do

REMEMBER

You have to move the decimal point the same number of places in both numbers.

Step 1. Set up the problem.

$.95\overline{)15.35}$

Step 2. Move the decimal point in the divisor two places to the right to make it a whole number.

$95.\overline{)15.35}$

Step 3. Move the decimal point in the dividend two places to the right. You will need to add a zero.

$95.\overline{)1535.}$

Step 4. Place the decimal point directly above in the answer. Divide as you would with whole numbers.

```
      16.15
95.)1535.00
     95
     585
     570
      150
       95
       550
       475
```

Round 16.15 to the nearest tenth. The ham weighed about 16.2 pounds.

Try These

Divide.

1. $.4\overline{)1.24}$ **2.** $.33\overline{)7.26}$ **3.** $.125\overline{)8.75}$

Divide.

4. $.38\overline{)5.7}$ **5.** $.6\overline{)7.2}$ **6.** $.36\overline{)2.592}$ **7.** $.21\overline{)20.58}$

8. $.61\overline{)44.53}$ **9.** $.47\overline{)7.52}$ **10.** $.11\overline{)1.166}$ **11.** $.56\overline{)21.28}$

12. $.9\overline{)1.89}$ **13.** $.323\overline{)277.78}$ **14.** $.23\overline{)9.43}$ **15.** $.8\overline{)98.4}$

Solving Problems

Solve.

16. Cassie purchased 60.5 pounds of mixed candy to be sold at the school bazaar. She divided the candy into .25 pound bags. How many bags of candy did she have to sell?

____

17. Enrique had to cut filler blocks .75 feet long out of a 2x4 piece of lumber that was 16.5 feet in length. How many blocks was he able to cut from the piece of lumber?

____

Check your answers on page 73.

LESSON

26 Dividing a Mixed Decimal by a Mixed Decimal

The Globe Potato Chip factory produces 4237.5 pounds of barbecue potato chips in one day. They are packed in 1.5-pound plastic bags. How many packages does the factory produce in one day?

Think

Divide the number of pounds by the amount in each package.

Do

Step 1. Set up the problem. Move the decimal point in the divisor one place to the right to make it a whole number.

$1.5\overline{)4237.5}$

Step 2. Move the decimal in the dividend the same number of places.

$15.\overline{)42375.}$

Step 3. Place the decimal point in the answer above the one in the dividend.

$15\overline{)42375.}$

Step 4. Divide as you would with whole numbers.

They package 2,825 bags of potato chips in one day.

```
      2825.
  15)42375.
     30
     123
     120
       37
       30
        75
```

Try These

Divide.

1. $1.75\overline{)22.75}$
2. $71.1\overline{)2.14722}$
3. $1.04\overline{)5.304}$

Divide.

4. $4.8\overline{)30.72}$

5. $2.29\overline{)393.88}$

6. $1.55\overline{)717.65}$

7. $1.03\overline{)9.476}$

8. $2.19\overline{)199.29}$

9. $1.6\overline{)30.72}$

10. $12.4\overline{)39.184}$

11. $1.51\overline{)21.14}$

12. $6.4\overline{)4.1664}$

Solving Problems

Solve.

13. Mia's 4-cylinder car used 9.1 gallons of a gas on a 205.75 mile trip. Find the number of miles per gallon to the nearest tenth.

____

14. Jana was hired to work at the hardware store for $4.35 per hour. She earned $26.10 her first day. How many hours did she work?

____

Check your answers on page 73.

LESSON

27 Dividing Money

Twenty-two members of the Sidney Adult Basic Education class divided the cost of a trip to the Denver Art Museum and the planetarium. The total cost was $402.60. What was each student's share of the cost?

Think

Divide the total cost by the number of students.

Do

Step 1. Set up the problem. The divisor is a whole number. Place the decimal point in the answer above the one in the dividend.

$$22\overline{)402.60}$$

Step 2. Divide as you would with whole numbers.

$$\begin{array}{r} 18.30 \\ 22\overline{)402.60} \\ \underline{22} \\ 182 \\ \underline{176} \\ 6\ 6 \\ \underline{6\ 6} \end{array}$$

Each student paid $18.30.

Try These

Divide.

1. $61\overline{)\$311.10}$ (answer: \$ ___ . ___)

2. $7\overline{)\$.49}$ (answer: \$ ___)

3. $10\overline{)\$10.71}$ (answer: \$ 1.071)

Since you only need two decimal places to show cents, round the answer to the hundredths place. $1.071 rounds to _____.

PRACTICE

Divide.

4. $120\overline{)\$43.20}$ (answer: \$)

5. $25\overline{)\$625.00}$

6. $46\overline{)\$400.20}$

7. $7\overline{)\$1484.00}$

8. $15\overline{)\$18.60}$

9. $29\overline{)\$1751.60}$

10. $31\overline{)\$15.50}$

11. $.75\overline{)\$2.25}$

12. $8\overline{)\$7.20}$

Solving Problems

Solve. Round your answers to the nearest cent.

13. Kurt was recently hired as an aide at the North Valley Convalescent Center. He will receive an annual salary of $14,726. What is his monthly salary?

____

14. Maria and Cheryl share an apartment and all expenses. If their utilities this month were $148.52, what was each person's share?

____

Check your answers on page 73.

LESSON

28 Dividing a Decimal by 10, 100 or 1,000

At his farm, Mr. Kuehn is planting a windbreak that is 950.5 feet long. He is planting Russian olive trees 10 feet apart. About how many trees will he need?

Think

Divide the length of the windbreak by the distance between the trees.

Do

Set up the problem. Divide.

$$\begin{array}{r} 95.05 \\ 10\overline{)950.5} \\ \underline{90} \\ 50 \\ \underline{50} \\ 50 \\ \underline{50} \end{array}$$

He will need about 95 trees.

Study these division facts:

$$950.5 \div 10 = 95.05$$

$$950.5 \div 100 = 9.505$$

$$950.5 \div 1000 = .9505$$

Do you see a pattern? To divide by 10, 100, or 1000, move the decimal point in the dividend to the left one place for every zero in the divisor.

Here's another example:

$$3.73 \div 100 = .0373$$

In this example it was necessary to add a zero in order to move the decimal point enough places.

Try These

1. $10\overline{)47.1}$ (answer shown: 4.)

2. $1000\overline{)3692.1}$

3. $100\overline{)6.48}$

PRACTICE

Place the decimal point correctly in the answer. Add zeros if necessary.

4. 125.3 ÷ 10 = 1253

5. .3629 ÷ 100 = 3629

6. 59.71 ÷ 100 = 5971

7. 2.1 ÷ 10 = 21

8. 63.9 ÷ 1000 = 639

9. 74.5 ÷ 10 = 745

10. .1 ÷ 10 = 1

11. 93.7 ÷ 10 = 937

12. .063 ÷ 10 = 63

13. 2.54 ÷ 100 = 254

14. 831 ÷ 10 = 831

15. 109.9 ÷ 1000 = 1099

16. 104.72 ÷ 1000 = 14072

17. 2.76 ÷ 10 = 276

Solving Problems

Solve.

18. Risa earned $38.50 in tips during the dinner rush. If she served 10 groups of people, how much was her average tip?

____

19. Jorge paid $35.00 to print 1,000 flyers. How much is he paying for each flyer?

____

Check your answers on page 74.

LESSON

29 Rounding to the Leading Digit in Division

Fifty-four senior citizens chartered a bus for a tour. The cost for transportation, which was shared equally by the travelers, was $1,693.40. About how much was each traveler's share to pay?

Think

The word *about* tells you to estimate an answer. To solve the problem, you need to divide the cost for transportation by the number of travelers. Before doing the work, round both numbers to the leading digit.

Do

Step 1. Round both numbers to the leading digit.

$1,693.40 rounds to $2,000.

54, the number of travelers, rounds to 50.

Step 2. Divide.

$$\begin{array}{r} \$40 \\ 50\overline{)\$2000} \\ \underline{200} \\ 0 \end{array}$$

Each traveler will pay about $40. The actual expense would be $31.36.

Try These

Round the numbers to the leading digits and divide.

1. $75\overline{)35.0}$ → $80\overline{)40.0}$

2. $19.2\overline{)42.2}$

3. $55\overline{).56}$

PRACTICE

Round the numbers to the leading digits and divide.

4. 51.2 ÷ 125 = _____

5. .51 ÷ 15 = _____

6. 6.42 ÷ 2.5 = _____

7. .95 ÷ 1.2 = _____

8. .125 ÷ .36 = _____

9. 12.5 ÷ .48 = _____

10. 7.5 ÷ 2.1 = _____

11. .365 ÷ .18 = _____

12. 651.9 ÷ 705 = _____

13. .89 ÷ .12 = _____

14. 2.99 ÷ 1.52 = _____

15. 225.8 ÷ 5.4 = _____

16. 6.1 ÷ 2.5 = _____

17. 13.9 ÷ 5.1 = _____

Solving Problems

Round the numbers to the leading digits and divide.

18. Miguel borrowed $18,940.60 to buy a new car. He plans on repaying the money by paying $280 a month. How many months will he need to repay the money?

____

19. Gao Min rented a compact car from the Clark car agency at a rate of $1.20 a mile. The total bill was $379.40. About how many miles was she charged for?

____

Check your answers on page 74.

LESSON

30 Decimal Division Review

Divide.

1. $17.20 ÷ 4 = _____
2. 1177.6 ÷ 512 = _____
3. 52.8 ÷ 10 = _____
4. 7.14 ÷ 4.2 = _____
5. .006 ÷ 5 = _____
6. 24.72 ÷ 2.17 = _____
7. 1.35 ÷ .5 = _____
8. .512 ÷ 32 = _____
9. 140.22 ÷ .6 = _____
10. 3.367 ÷ .91 = _____
11. 52.8 ÷ 100 = _____
12. 6.106 ÷ 86 = _____
13. 1.2 ÷ .08 = _____
14. 10.125 ÷ .81 = _____
15. .81 ÷ 9 = _____
16. 16.016 ÷ .16 = _____
17. 90 ÷ .15 = _____
18. 185.64 ÷ 39 = _____
19. .246 ÷ 6 = _____

Divide and round the answer to the nearest tenth.

20. 24 ÷ 36 = _____
21. 140.94 ÷ 243 = _____
22. 7 ÷ .30 = _____

Divide and round the answer to the nearest hundredth.

23. 44.8 ÷ 37 = _____
24. 66.34 ÷ .62 = _____
25. 5.2 ÷ 29 = _____

Solving Problems

Solve.

26. The salmon that Tyson caught weighed 9.4 pounds. After filleting it, he divided the salmon into 3 packages. How much did each package weigh?

____

27. Alicia spent $18.75 for school lunches for her three children for the week. How much does it cost for one child to buy school lunch for the week?

____

28. The Park La Brea Hornets won 43 of their first 75 games of the season. Find their winning percentage by dividing the games won by the games played. Round to the thousandths place.

____

Check your answers on page 74.

LESSON

31 Review of Multiplication and Division of Decimals

Perform the indicated operation.

1. 612 ÷ .04 = _____
2. 28.4 ÷ 100 = _____
3. 1 ÷ .5 = _____
4. 81.19 ÷ 23 = _____
5. 101.86 ÷ .463 = _____
6. .45 × .89 = _____
7. 5.75 × .04 = _____
8. 273.672 ÷ 36.2 = _____
9. .045 × 10 = _____
10. .51 × 100 = _____
11. 1036 ÷ .037 = _____
12. 9.2 × .34 = _____
13. 9.2 × 1.08 = _____
14. 42.42 ÷ 10.1 = _____
15. 8.45 ÷ 05 = _____
16. 18.607 ÷ 8.09 = _____
17. .01 ÷ .5 = _____
18. 381.17 ÷ 81.1 = _____
19. .24 × 137 = _____
20. .288 ÷ 15 = _____

Select the best answer for each of these examples.

	a)	b)	c)
21. 255 ÷ .5 =	51.0	500	510
22. .78 × .1 =	.78	.01	.078
23. 43.6 ÷ 100 =	.436	436	4360
24. 51 ÷ .3 =	1.7	170	17
25. 98 ÷ .6 =	163.	163.3	160

26. Wanitta Transfer pays its over-the-road drivers $.60 per mile and allows $45 per day more for motels and food. How much will Kyle receive for a five-day trip if he averages 453 miles per day?

____

27. Tomás drives for the same company but only on short hauls. He works 48 hours a week and is paid 1.5 overtime for all hours over 40 hours. His base pay is $11.00 an hour. What are his weekly wages?

____

28. The odometer reading is 53241.4 miles at the beginning of a trip and 56328.0 miles at the end of the trip. Find the number of miles driven per gallon if 140.3 gallons of gasoline were used on the trip.

____

29. Keo receives $13,488 each year from his Social Security Retirement fund. How much does he receive each month?

____

30. Last year Reyes Prescriptions had 500 charge account customers. They project an increase of approximately 1.75 this year. How many more customers do they expect to have?

____

31. The U.S. toy market (not including video games) accounted for $17.5 billion in sales last year. Action figures represented .05 of the total sales. How much money was spent on action figures?

____

32. The first U.S. telegraph transmission lines were 40 miles long. The cost of the lines was $30,000. What was the cost per mile?

____

33. Babbitt Nursery developed a new rose. "Sweet Dream" is available for the first time this year. Five rose bushes sell for $27.75. What is the price for one rose bush?

____

34. At the Bonneville Salt Flats, Eric Chin drove his car 1,320 feet in 4.35 seconds. What is his rate in feet per second?

____

35. Water users in Sidney pay .098 cents for each gallon of water used. How much would a customer pay for 1,000 gallons of water usage?

____

Check your answers on page 74.

Appendix: Multiplication Facts

Multiplication Table

1	2	3	4	5	6	7	8	9	10	11	12
2	4	6	8	10	12	14	16	18	20	22	24
3	6	9	12	15	18	21	24	27	30	33	36
4	8	12	16	20	24	28	32	36	40	44	48
5	10	15	20	25	30	35	40	45	50	55	60
6	12	18	24	30	36	42	48	54	60	66	72
7	14	21	28	35	42	49	56	63	70	77	84
8	16	24	32	40	48	56	64	72	80	88	96
9	18	27	36	45	54	63	72	81	90	99	108
10	20	30	40	50	60	70	80	90	100	110	120
11	22	33	44	55	66	77	88	99	110	121	132
12	24	36	48	60	72	84	96	108	120	132	144

This table shows the products of the numbers 1 to 12 multiplied by any number from 1 to 12.

$4 \times 3 =$ Find the number 4 in the first column and the number 3 in the top row. Then follow down and across in straight lines until the two lines intersect. The number at the intersection point is the answer. $4 \times 3 = 12$

Answer Key

Lesson 1 → MULTIPLYING WHOLE NUMBERS (PAGE 2)

1. $12,305.00 **2.** $13,650.00 **3.** 6,624 **4.** 69,564 **5.** 115,668

6. 102,992 **7.** 74,290 **8.** 875,235 **9.** 456,344 **10.** 318,008

11. 526,500 **12.** 401,280 **13.** 783,702 **14.** 402,696 **15.** 1,339,690

16. 475,797 **17.** 93,449 **18.** 4,144 squares **19.** 26,780 squares

Lesson 2 → ROUNDING TO THE NEAREST WHOLE NUMBER (PAGES 3–4)

1. 524 **2.** 371 **3.** 23 **4.** 3 **5.** 147

6. 67 **7.** 564 **8.** 882 **9.** 1,342 **10.** 369

11. 7,400 **12.** 49 **13.** 26 **14.** 2,488 **15.** 165, 195, 184

16. Possible answers are: 28.1, 27.7. Answers will vary.

17. Possible answers are: 341.3, 340.6. Answers will vary.

Lesson 3 → ROUNDING OFF TO VARIOUS DECIMAL PLACES (PAGES 5–6)

1. 257.8 **2.** 496.37 **3.** 473.25 **4.** 56.4 **5.** 1,235.1 **6.** 581.6

7. 965 **8.** 676.3 **9.** 47.562 **10.** 300 **11.** 2.0 **12.** .9

13. .7, .7, .7
All the percentages are equal.

14. .73, .68, .68
John Madden has the highest percentage.
Joe Gibbs and George Allen have equal percentages.

15. .3, .3

Lesson 4 → SETTING UP MULTIPLICATION PROBLEMS (PAGES 7–8)

1. 3 1392 × .871

2. 5 .5831 × 5.9

3. 5 1.392 × 1.51

4. 4 1.79 × .52

5. 5 .427 × .81

6. 4 .385 × 2.9

7. 4 11.074 × .8

8. 5 .917 × .18

9. 4 1.01 × .97

10. 4 6.84 × 3.98

11. "A" is easier to multiply.
The answers are the same.

12. 4 $1.69 × 1.57

Lesson 5 → MULTIPLYING A WHOLE NUMBER BY TENTHS (PAGES 9–10)

1. 2.8 **2.** 2.7 **3.** .7 **4.** 0 **5.** 1.8 **6.** 3.2 **7.** 5.4 **8.** .6 **9.** 6.3

10. 1.5 **11.** 0 **12.** 2.5 **13.** 4.2 **14.** 4 **15.** 3.6 **16.** 2.1 **17.** 5.4 **18.** 8.1

19. 2.4 miles **20.** 1.2 centimeters **21.** 3.2 gallons **22.** 2.7 pounds

Lesson 6 → MULTIPLYING A WHOLE NUMBER BY HUNDREDTHS (PAGES 11–12)

1. 5.67 **2.** 13.95 **3.** 15.36 **4.** 4.18 **5.** 81.27 **6.** 92.48
7. 183.12 **8.** 804.42 **9.** 34.81 **10.** 80.94 **11.** 140.91 **12.** 118.08
13. 402.57 **14.** 561.09 **15.** 170.14 **16.** 131.44 **17.** 116.03 **18.** 678.9
19. 596.82 **20.** 109.12 **21.** 9 pounds **22.** 6,820 people **23.** 285 people

Lesson 7 → MULTIPLYING A WHOLE NUMBER BY THOUSANDTHS (PAGES 13–14)

1. 4.935 **2.** 5.808 **3.** 6.916 **4.** 55.473 **5.** 95.734 **6.** 630.252
7. 232.971 **8.** 169.824 **9.** 10.138 **10.** 22.968 **11.** 9.99 **12.** 72.324

13. 464.916 miles
35.084 miles to complete race
$154.972 \times 3 = 464.916$
$500 - 464.916 = 35.084$

14. 382.977 miles

Lesson 8 → MULTIPLYING A WHOLE NUMBER BY A DECIMAL (PAGES 15–16)

1. 2.5 **2.** .306 **3.** 12.9 **4.** 10.5 **5.** 7.8 **6.** .2 **7.** 18.2 **8.** 9.9
9. 5.05 **10.** 1.64 **11.** .69 **12.** 1.212 **13.** .228 **14.** .96 **15.** 4.6 **16.** 22.4 miles
17. 144.6 kilograms **18.** 26,752 pounds **19.** 2 lamps **20.** 19,600 women

Lesson 9 → MULTIPLY A DECIMAL BY A DECIMAL (PAGES 17–18)

1. 1.305 **2.** 25.992 **3.** 4.3848 **4.** .2697 **5.** 2.1726 **6.** .2343
7. 19.1064 **8.** 17.318 **9.** 22.94 **10.** 1.36 **11.** .39216 **12.** 4.2036
13. 3.822 **14.** 6.24 **15.** 565.32 **16.** 4.819 **17.** 26.884 **18.** .31155
19. .4152 **20.** 2.7924 **21.** .0946 liters **22.** 8.96 ounces

Lesson 10 → ZERO AS A PLACE HOLDER (PAGES 19–20)

1. .02485 **2.** .00063 **3.** .44772 **4.** .06045 **5.** .04524 **6.** .027296
7. .00126 **8.** .0093 **9.** .084546 **10.** .035112 **11.** .0104 **12.** .0468
13. .039273 **14.** .0837 **15.** .0144 **16.** .035 **17.** .08

Lesson 11 → MULTIPLYING MONEY (PAGES 21–22)

1. $244.12 **2.** $97.76 **3.** $248.22 **4.** $138.06 **5.** $124.52 **6.** $132.60
7. $857.61 **8.** $113.22 **9.** $162.68 **10.** $433.16 **11.** $151.56

12. **AUTO SERV Order**

Quantity	Description	Unit Cost	Total
5	Ozark Motor Oil	1.15	$ 5.75
1	Ozark Oil Filter	3.78	$ 3.78
6	Better Spark Plugs	2.73	$16.38
2	Barr's Lube (Tube)	1.98	$ 3.96
2	Ozark Coolant	6.94	$13.88
		Total	$43.75

13. $4.38

Lesson 12 → MULTIPLYING SMALL DECIMAL NUMBERS (PAGES 23–24)

1. .0075 **2.** .026 **3.** .02697 **4.** .024 **5.** .0042 **6.** .741 **7.** .14 **8.** .0495

9. .567 **10.** 6.3 **11.** .005 **12.** .3528 **13.** .493 **14.** .00923 **15.** .0836 **16.** .918

17. 750 lbs $2000 \times .75 = 1500$
$1500 \times .5 = 750$

18. .195 inch **19.** .26 **20.** .04

Lesson 13 → MULTIPLYING A DECIMAL BY 10, 100 AND 1,000 (PAGES 25–26)

1. .45 **2.** 178 **3.** 239 **4.** 517 **5.** 53 **6.** 17,890 **7.** 612

8. 340 **9.** 24,710 **10.** 160 **11.** .67 **12.** 151.7 **13.** 80.3 **14.** 2,357

15. $100 $\$10 \times 10 = 100$
$75 $\$7.50 \times 10 = 75$

16. 4,380 pounds

Lesson 14 → ROUNDING AND MULTIPLYING DECIMALS (PAGES 27–28)

1. 1.2 **2.** .2; $7 \times .2 = 1.4$ **3.** 5 $5 \times 1 = 5$ **4. a)** $.2 \times 1$

5. a) $8 \times .3$ **6. b)** 11×1 **7. c)** $179 \times .5$ **8. a)** 96×1.7

9. b) $9,400 \times 21$ **10.** 295 $59 \times 5 = 295$ **11.** 12,220 $611 \times 20 = 12,220$

12. 2 $1 \times 2 = 2$ **13.** 1,001 $13 \times 77 = 1,001$ **14.** 470 $47 \times 10 = 470$

15. 8 $4 \times 2 = 8$ **16.** 7,938 pounds $147 \times 54 = 7,938$ **17.** $2,100 $7,000 \times .3 = 2,100$

Lesson 15 ROUNDING TO THE LEADING DIGIT IN MULTIPLICATION (PAGES 29–30)

1. .08 $.4 \times .2 = .08$ **2.** 45 $50 \times .9 = 45$ **3.** 60 $10 \times 6 = 60$
4. .56 $.8 \times .7 = .56$ **5.** .8 $2 \times .4 = .8$ **6.** 30 $300 \times .1 = 30$
7. 540 $90 \times 6 = 540$ **8.** 1,800 $600 \times 3 = 1{,}800$ **9.** 30 $600 \times .05 = 30$
10. 4.4 $44 \times .1 = 4.4$ **11.** .18 $.3 \times .6 = .18$ **12.** 2,000 $1{,}000 \times 2 = 2{,}000$
13. 3,000 $5{,}000 \times .6 = 3{,}000$ **14.** 9,000 $900 \times 10 = 9{,}000$ **15.** 56 $80 \times .7 = 56$
16. $4.80, $4.67 **17.** $90, $75.63 **18.** 10 million, 5.765 million

Lesson 16 REVIEW OF DECIMAL MULTIPLICATION (PAGES 31–32)

1. $2,550 **2.** .08 **3.** .072 **4.** 125 **5.** 552.72 **6.** .0009
7. $10,750 **8.** 2.7 **9.** 1.52028 **10.** .002364 **11.** 102.03 **12.** 576.8
13. 16.2268 **14.** 7,155.2 **15.** 3,996.72 **16.** 3,739.518 **17.** .0875 **18.** 74.3461
19. 1.4076 **20.** .6524 **21.** $791.70 **22.** $49 **23.** $117 **24.** 97,864

Lesson 17 REVIEW OF DIVISION OF WHOLE NUMBERS (PAGES 33–34)

1. 242 **2.** 1,620 **3.** 49 **4.** 601 **5.** 41 R10 **6.** 72 R65 **7.** 18 R3
8. 20 R136 **9.** 150 R8 **10.** 73 **11.** 35 **12.** 128 **13.** 17 **14.** 350 miles

Lesson 18 DIVIDING A DECIMAL BY A WHOLE NUMBER (PAGES 35–37)

1. 2.6 **2.** 3.7 **3.** 1.2475 **4.** .102 **5.** .35 **6.** 17.7 **7.** .46 **8.** .44
9. 3.6 **10.** .055 **11.** 1.2 **12.** .06 **13.** 11.1 **14.** .072 **15.** .61 **16.** 12 trucks
17. Yes. The twelfth truck will carry a part load. **18.** $44.12 **19.** 55.1 miles **20.** $19.95

Lesson 19 ZERO AS A PLACE HOLDER IN DIVISION (PAGES 38–40)

1. .091 **2.** .072 **3.** .0091 **4.** .04 **5.** .05 **6.** .031
7. .0501 **8.** .09 **9.** .073 **10.** .019 **11.** .17 **12.** .108
13. .019 **14.** .0081 **15.** .0059672 **16.** $3.05 **17.** $11.06 **18.** 4.02 feet

Lesson 20 DIVIDING A WHOLE NUMBER BY A 10TH, 100TH, OR 1,000TH (PAGES 41–42)

1. 700 **2.** 42,000 **3.** 3,200 **4.** 90 **5.** 900
6. 74,000 **7.** 5,400 **8.** 11,000 **9.** 650 **10.** 12,100
11. 26,000 **12.** 10 **13.** 93,000 **14.** 4,110 **15.** 36,400
16. 701,000 **17.** 1,000 **18.** 240 **19.** 60 students **20.** 2,900 employees

Lesson 21 ······▶ DIVIDING A WHOLE NUMBER BY A DECIMAL (PAGES 43–45)

1. 50 **2.** 100 **3.** 600 **4.** 400 **5.** 50 **6.** 6,000 **7.** 170 **8.** 1,700
9. 21,300 **10.** 6,700 **11.** 8,500 **12.** 1,030 **13.** 86,000 **14.** 5,500
15. 400 **16.** 40 **17.** 20 **18.** 10 **19.** 48

Lesson 22 ······▶ DIVIDING A WHOLE NUMBER BY A LARGER WHOLE NUMBER (PAGES 46–47)

1. .75 **2.** .875 **3.** .666 .67 **4.** .41 **5.** .61 **6.** .05 **7.** .63
8. .22 **9.** .13 **10.** .393 **11.** .048 **12.** .808 **13.** .333 **14.** .481

Lesson 23 ······▶ DIVIDING A DECIMAL BY A DECIMAL (PAGES 48–49)

1. 4 **2.** .8 **3.** .23 **4.** .98 **5.** .76 **6.** 1.3 **7.** 11 **8.** .9 **9.** .105
10. .83 **11.** 1.1 **12.** .7 **13.** .75 **14.** 103.6 **15.** .46 **16.** 23 **17.** 80 No

Lesson 24 ······▶ DIVIDING A MIXED DECIMAL BY A WHOLE NUMBER (PAGES 50–52)

1. .003 **2.** .12 **3.** 0.2937717 **4.** .8 **5.** .2 **6.** .2 **7.** 2.6 **8.** .8
9. .5 **10.** .17 **11.** .11 **12.** .31 **13.** .07 **14.** .17 **15.** .24 **16.** .638
17. .431 **18.** 4.2 pounds **19.** 6.7 yards

Lesson 25 ······▶ DIVIDING A MIXED DECIMAL BY A DECIMAL (PAGES 53–54)

1. 3.1 **2.** 22 **3.** 70 **4.** 15 **5.** 12 **6.** 7.2 **7.** 98 **8.** 73 **9.** 16
10. 10.6 **11.** 38 **12.** 2.1 **13.** 860 **14.** 41 **15.** 123 **16.** 242 **17.** 22

Lesson 26 ······▶ DIVIDING A MIXED DECIMAL BY A MIXED DECIMAL (PAGES 55–56)

1. 13 **2.** .0302 **3.** 5.1 **4.** 6.4 **5.** 172 **6.** 463 **7.** 9.2
8. 91 **9.** 19.2 **10.** 3.16 **11.** 14 **12.** .651 **13.** 22.6 miles per gallon **14.** 6

Lesson 27 ······▶ DIVIDING MONEY (PAGES 57–58)

1. $5.10 **2.** $.07 **3.** $1.07 **4.** $.36 **5.** $25 **6.** $8.70
7. $212 **8.** $1.24 **9.** $60.40 **10.** $.50 **11.** $3 **12.** $.90
13. $1,227.17 **14.** $74.26

Lesson 28 → DIVIDING A DECIMAL BY 10, 100 OR 1,000 (PAGE 60)

1. 4.71 **2.** 3.6921 **3.** .0648 **4.** 12.53 **5.** .003629 **6.** .5971 **7.** .21
8. .0639 **9.** 7.45 **10.** .01 **11.** 9.37 **12.** .0063 **13.** .0254 **14.** 83.1
15. .1099 **16.** .10472 **17.** .276 **18.** $3.85 **19.** $.035

Lesson 29 → ROUNDING TO THE LEADING DIGIT IN DIVISION (PAGES 61–62)

1. .5 **2.** 2 **3.** .01 **4.** .025 **5.** .05 **6.** 2 **7.** 1 **8.** .25 **9.** 20 **10.** 4
11. 2 **12.** 1 **13.** 9 **14.** 1.5 **15.** 40 **16.** 2 **17.** 2 **18.** 66.67 **19.** 400

Lesson 30 → DECIMAL DIVISION REVIEW (PAGES 63–64)

1. $4.30 **2.** 2.3 **3.** 5.28 **4.** 1.7 **5.** .0012 **6.** 11.392 **7.** 2.7
8. .016 **9.** 233.7 **10.** 3.7 **11.** .528 **12.** .071 **13.** 15 **14.** 12.5
15. .09 **16.** 100.1 **17.** 600 **18.** 4.76 **19.** .041 **20.** .7 **21.** .6
22. 23.3 **23.** 1.21 **24.** 107 **25.** .18 **26.** 3.1 pounds **27.** $6.25 **28.** .573

Lesson 31 → REVIEW OF MULTIPLICATION AND DIVISION OF DECIMALS (PAGES 65–67)

1. 15,300 **2.** .284 **3.** 2 **4.** 3.53 **5.** 220 **6.** .4005 **7.** .23
8. 7.56 **9.** .45 **10.** 51 **11.** 28,000 **12.** 3.128 **13.** 9.936 **14.** 4.2
15. 169 **16.** 2.3 **17.** .02 **18.** 4.7 **19.** 32.88 **20.** .0192
21. **c)** 510 **22.** **c)** .078 **23.** **a)** .436 **24.** **b)** 170 **25.** **b)** 163.3

26. $496.80 45 × 5 = 225
.60 × 453 = 271.80
225 + 271.80 = 496.80

27. $572 40 × 11 = 440
11 × 1.5 = 16.50
8 × 16.50 = 132
440 + 132 = 572

28. 22 56,328 – 53,241.4 = 3,086.6
3,086 ÷ 140.3 = 22

29. $1,124 13,488 ÷ 12 = 1,124

30. 375 500 × 1.75 = 875
875 – 500 = 375

31. .875 billion 17.5 × .05 = .875

32. $750 30,000 ÷ 40 = 750

33. $5.55 27.75 ÷ 5 = 5.55

34. 303.45 ft/sec 1,320 ÷ 4.35 = 303.44827

35. $98 .098 × 1,000 = 98

NOTES

NOTES